UNDERSTANDING
WASTE = BIOENERGY

UNDERSTANDING WASTE = BIOENERGY

Dr. Smarajit Roy

authorHOUSE®

AuthorHouse™
1663 Liberty Drive
Bloomington, IN 47403
www.authorhouse.com
Phone: 1-800-839-8640

First published by AuthorHouse   06/01/2011

ISBN: 978-1-4567-8354-9 (sc)

Printed in the United States of America

This book is printed on acid-free paper.

CONTENTS

ABSTRACT

Biomass, a renewable energy source, is biological material from living, or recently living organisms, such as wood, waste, (hydrogen) gas, and alcohol fuels. Biomass is commonly plant matter grown to generate electricity or produce heat. Natural plants or cultivated plants (plants for food) forest residues (dead trees, branches and tree stumps), yard clippings, wood chips and garbage are often used for this. However, biomass also includes plant or animal body matter or animal waste and manure. Swage sludge has been used in producing of biogas from as early as 1859. Sewage sludge is the most common biomass for producing combined heat and power in developing countries. Biomass may also include biodegradable wastes that can be burnt as fuel (municipality waste incineration) to produce steam for communal heating. It excludes organic materials such as fossil fuels which have been transformed by geological processes into substances such as coal or petroleum.

However biomass is an energy source which is a stable alternative to petroleum. In addition to transportation fuel, biomass can be used to generate electricity in rural areas. The UK has responded to the EU Renewable Energy Directive by committing to generate15% of its energy requirement from renewable sources by 2020. While 1st and second generation of biomass is an important feedstock, there are also emerging technologies which make use of municipal, commercial, industrial waste stream will supplement or replace petroleum and diesel derivatives.

Recent developments use cultured bacteria that can transfer electrons to the surface of electrodes and making full use of fuel cell technology. Some scientists have developed other strain of bacteria that can convert rubbish in to biofuels that can be used as fuels for cars and trucks. Algae (sea weed) which grows abundantly in any marshy land, sea and river banks can used to used as renewable diesel, petroleum for cars, trucks for commercial and military aviation fuel.

KEYWORDS

Bio-Energy, Bio-Fuel, Advanced Bio-fuel, Municipality waste to Bio-fuel, Renewable Energy, New Bio-fuel products, Algae to bio-gas, Non food biomass, What is bio-gas?

INTRODUCTION

Different biotechnology solutions can potentially be combined to create "ecosystems" in which we use natural inputs including materials discarded or natural waste, expend minimum amount of energy and do not produce any waste. Biotechnology solutions in general should produce renewable energy which is affordable, sustainable with economic and environmental benefits. Biotechnology plants should consider infrastructure required including the availability of clean water requirements, quality and quantity of feedstock availability, feed stock pretreatment process requirements, disposal of waste and emission regulation requirements.

Most of us would bet the next real economic boom will belong to renewable energies. And, it still seems that despite mounting evidence societies will only adjust their behavior significantly in the face of huge catastrophic events, seemingly unable to prevent them beforehand. Our policy leaders need to switch from a perception of crisis, to one of opportunity. While mitigating climate change, these industries already demonstrate their ability to substitute the fossil and nuclear energy base, thereby creating jobs and improving energy security, as well as welfare.

Petroleum and coal have long been the cheap dates at the energy ball. Now, suitors are starting to swing to a tune of bioenergy. In April 2010 the Technology Entertainment and Design (TED) talk, U.S. Navy set aggressive targets to reduce dependence on foreign oil. At this TED talk on energy, Bill Gates said, "We need solutions, either one or several, that have unbelievable scale and unbelievable liability." President Obama said in his 2010 State of the Union address, "The nation that leads the clean energy economy will be the nation that leads the global economy." The need for clean, renewable energy will not vanish at midnight; indeed, it will increase every year.

MAIN EXPOSITION

Origin of Biogas

The first anaerobic digester was built by a leper colony in Bombay, India in 1859. In 1895 the technology was developed in Exeter, England, where a septic tank was used to generate gas for the sewer gas destructor lamp, a type of gas lighting.

Tap Water and Sanitation has developed in convenient partnership. The importance of the word *"Convenienence"* is significant in this present context. This is what you enquire of when you plan a journey. Sanitation is a nasty and embarrassing problem but a necessity which is difficult to get rid off.

The water on tap was invented a few hundred years ago. Flush toilet was invented about 150 years ago. But the modern version was invented by Thomas Crapper just over 100 years ago. His invention of symphonic flush, which pulls the water out of the bowl and in to the pipe, the *"S"* bend was god sent. It performs two tasks, prevents odor emerging from *"Crap!"* and eliminate *"Crap"* sticking to the basin. Since then have never looked back. Now it's status symbol. You must have a Toilet in your bedroom!

The sewer system may either be a gravity sewer or a vacuum sewer. Gravity sewers allow the wastewater to flow from the properties to our treatment works via a number of pumping stations which pumps the wastewater to the next section of gravity sewer.

In a vacuum sewer the wastewater is collected in underground 'pots' near customer's properties. A vacuum station then creates a vacuum which effectively sucks the wastewater from the pot to the pumping station where it is pumped to the works.

During the last twenty years people realized the flaw in the system. As you would appreciate the system was designed by civil engineers and its now time to look back and analyze from the prospective of sustainable supply chain management issues. This gives us an opportunity to review processes and operational concepts from a different perspective. It should incorporate the role of the environment in supply chain value creation; leading companies are finding unique value opportunity in sustainable supply change management. I would provide you with ten case studies in this presentation.

After we flush, our waste ends up in either the air, the water, or in the ground. Organic materials are converted by bacteria into carbon dioxide and nutrients in the soil. At wastewater treatment plants, the inorganic solids are screened out and ultimately incinerated or buried underground at a landfill. Most organic solids are converted by bacteria into gas (especially N_2 and CO_2) and vented into the atmosphere, or into a nearby river or sea. The biosolids can be buried in a landfill, but that's expensive. Biosolids can be incinerated so a smaller volume of ash which can be buried in a landfill, but that's still expensive. The solution is to treat the biosolids so they pose no health risk, and then use them as fertilizer. About 60 percent of sewage sludge in the United States was used for land application. Sludge must be processed further before spreading to eliminate most trace metals, such as arsenic, lead, mercury, and pathogens, including viruses and bacteria.

Basic Anaerobic Digestion Process

Anaerobic digestion is an energy process that's most simply described as what happens in a cow's stomach. Feed can be broken down into methane and carbon dioxide gas. Naturally occurring bacteria, in the absence of oxygen, is able to break down that solid matter into biogas.

Benefits of Anaerobic Digestion Technologies

There are a multitude of benefits:

- Creating and using clean, low cost renewable energy
- Replace fossil fuel
- Reduce methane emission from land fills
- Displacing chemical fertilizers
- Reduce energy footprint of waste treatment plants. Plants are located in close proximity.
- Increase energy independence
- Create Regional jobs in agriculture, supply logistics, engineering, plant construction and Maintenance
- A variety of organic materials can be used in combination with others.
- Biomethane is more flexible in application than any other renewable source of energy. Its ability to be injected directly into the existing natural gas grid.

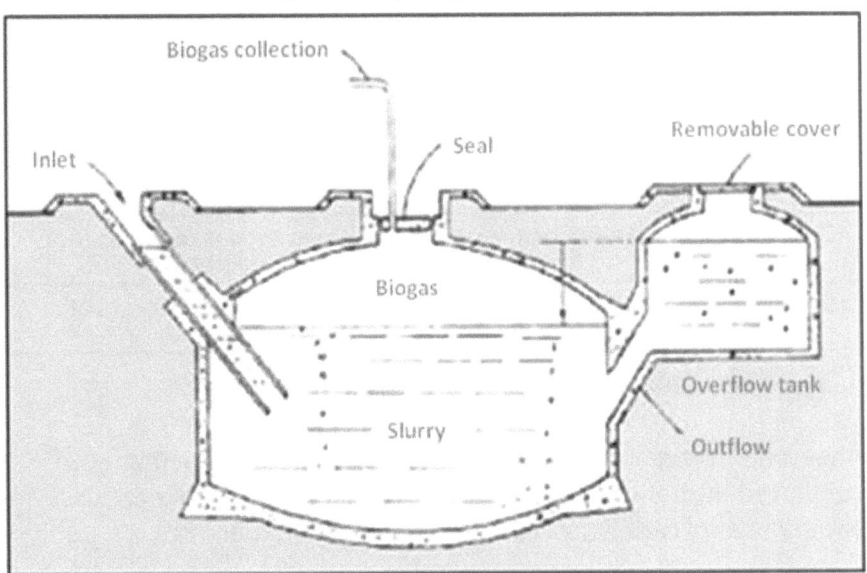

Fixed Dome Digester

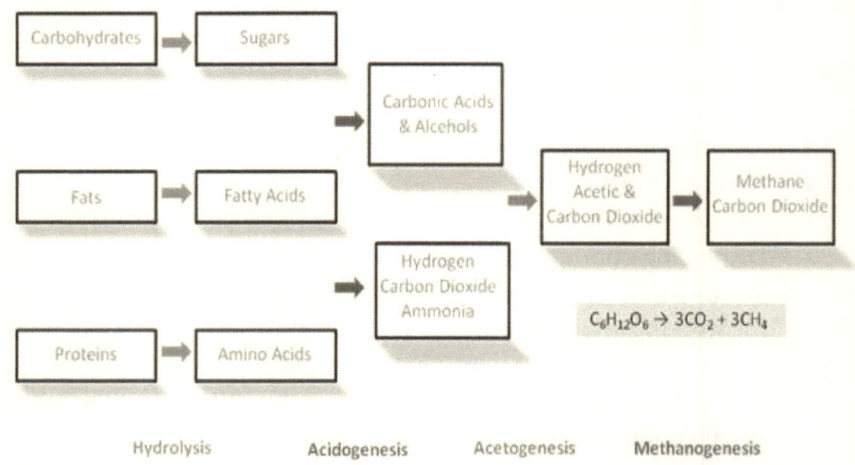

Anaerobic Digestion Process

Anaerobic digestion technologies produce energy that can be converted into electricity, heat, and natural gas suitable for use in a variety of applications, including transportation fuel.

HYDROLYSIS	Break downs the carbohydrates to Monomers, sugars and amino acids
ACIDOGENESIS	Monomers are transformed to fatty acids, similar to the way milk sours
ACETOGENESIS	Fatty acids are broken down to weaker acids_Acetic Acids, Carbon Dioxide and Hydrogen
METHANOGENESIS	Bacteria consumes the hydrogen and turn acetic acid in to methane and carbon dioxide

Table 1: Anaerobic Digestion Technologies

Anaerobic digestion technologies produce energy that can be converted into electricity, heat, and natural gas suitable for use in a variety of applications, including transportation fuel.

Combined Heat and Power (CHP)—the biogas is used in a combined heat and power unit to produce electricity and heat.

Pipeline-grade natural gas—the biogas is cleaned to a methane content of 97% and sold into existing natural gas pipelines.

Compressed natural gas (CNG)—the biogas is cleaned to a methane content of 98% and compressed to 3000 psi to be used as a transportation fuel.

State-of-the-art anaerobic digesters reduce the waste's volume, and their high temperature helps purify it, making it less bio-reactive and better fertilizer. By repurposing biogas as fuel, you are avoiding raw methane releases which are 23 times the climate warming potential of carbon dioxide.

Waste Water Treatment

Treatment includes removing solids and materials from the wastewater. Most plants are currently using a gaseous disinfectant method for this treatment. Metered chlorine gas enters the wastewater to achieve chlorination over a monitored contact time. A sulfur dioxide gas is then introduced to remove the chlorine before discharge occurs.

The new system uses liquid sodium hypochlorite solution (SHS) into the wastewater to achieve the required chlorination, followed by a liquid sodium bisulphate solution (SBS) to remove the chlorine.

Advanced Digestion Process

Thermophilic Digestion

Largest plant using this process is at the Anacis Island Wastewater Treatment Plant near Vancouver, Canada. This was designed to treat sludge from a population of 1,000.000 million people.

The process known as extended thermophilic digestion. The sludge travels in a continuous flow through tanks in series, ensuring that pathogens do not break through to the effluent biosolids as a result of short-circuiting through a single tank.

The plant can treat 480 million liters of wastewater per day, and is capable of being expanded to twice that capacity. Whereas in other jurisdictions effluent criteria are based on monthly averages, the Anacis plant has to meet daily maximum effluent criteria. Therefore it must perform optimally every day, which is a challenge for a plant that handles industrial wastewater and combined sewer flows in addition to domestic wastewater.

Methane gas produced from the sludge digestion is used to generate power and heat for the plant. The biogas fuels 4 Nos. 850 kW cogeneration engines, providing about 50 per cent of the plant power demand. Waste heat from the clean-burning co-generation engines is also recovered for space and process heating.

The plant-wide system communicates with more than 5,000 pieces of equipment, including more than 4,000 instruments, 700 motors, 12 programmable logic controllers, 50 electrical distribution system power monitors, 50 feeder protection relay and ground alarm detectors and more than 100 variable frequency drives.

WELtec BioPower

This is a biogas system is the result of finely tuned technologies. WELtec uses plants with stainless steel fermenters containing combination of long arm and submersible agitators. The material has to be heated to around 38 degrees Celsius, because this is the optimum temperature for the bacteria to thrive. The resulting biogas contains 52 percent methane, nearly 40 percent carbon dioxide and some other gases.

The feed stock could be waste from commercial kitchens, expired foods from supermarkets, waste fruit or old cooking oil. Wastes from slaughter house can also be used. Cyprus is well-known for its pig farming industry and is using this technology. The concept of the WELtec plant is to be established in Wuxi, a city of 4.5 million.

There are two main pre-digestion processes used in AAD in the UK

The Enzyme Hydrolysis	Monsal Process
The Thermal Hydrolysis	Cambi Process

Table 2: Pre-Digestion Processes

The water sector has 220 anaerobic digestion sites in the UK, feeding 88MW of generation capacity that produces up to 0.77TWh a year. This is expected to reach 115MW generating up to 1TWh by the end of the current investment period in 2010. This represents about 90 per cent of the total energy produced by anaerobic digestion across the UK. The other 10 per cent comes from agriculture, food processing and any other industry that produces biodegradable waste.

Enzymic Hydrolysis

With enzyme hydrolysis at 42 to 55 degree centigrade over several days, the result is greater conversion of organic matter into biogas 50 % reduction in sludge volume Biogas/CHP derived energy is produced.

Monsal Advanced Biological Hydrolysis

The Monsal Enzymic Hydrolysis process utilizes multiple CSTRs (continuous stirred-tank reactors) in series to harness the benefits of plug flow batch treatment without short circuiting. An advantage of the Monsal Enzymic Process over Acid Phase Digestion is that in using multiple tanks a hydrolysis profile across the reactors develops making the configuration more robust when treating variable sludge loads.

Each reactor vessel is mixed using gas mixing, and sludge is moved through the plant in reverse cascade batch, via high and low—level gas lifts. The first three reactor vessels are operated as serial mesophilic reactors at 42 degree centigrade for 1.5 days. The second three reactor vessels are operated as a serial-batch pasteurization stage at 55 degree centigrade with a retention time of 1.5 days.

Enhanced Enzymic Hydrolysis (Steam) Energy Balance

The design of the plant is for stage one to operate at 42 degree centigrade and stage 2 at 55 degree centigrade., with a hydraulic retention time of 1.5 to 2 days to allow sufficient but not excessive generation of volatile fatty acids (VFAs).

The design of the plant is for stage one to operate at 42 degree centigrade and stage 2 at 55 degree centigrade., with a hydraulic retention time of 1.5 to 2 days to allow sufficient but not excessive generation of volatile fatty acids (VFAs).

Energy Production at Avonmouth

The hydrolysis is catalyzed by enzymes, anaerobic bacteria. Without these enzymes, the hydrolysis would be impractically slow under the conditions found in mesophilic anaerobic digestion (MAD).

Phased digestion is a separation process of hydrolysis stage from the acidogenesis and methanogenesis stages to optimize the process.

The thermal hydrolysis process breaks down the biosolids, reducing the volume by about 40 per cent and produces biogas which is used to generate electricity. Because the volume of biosolids reaming low, there is less to dispose.

The biosolids process is environmentally friendly compared to traditional methods. Wastewater are collected and treated from 5.5 million customers. The centre is designed to treat up to 19,000 TDS/annum.

The enhanced treated sludge achieves a 6 log reduction in E.coli concentration, an absence of Salmonella and cake dry solids above 24% w/w. Upgrading was done at a cost of £27.5 million.

Sludge Load	30,700 tds/annum
Power Generation	34 Mwe
EH HRT	3 days
Gas Yield	425 Cubic meter / tds
Utilization	100 %
Supplementary	Fuel 0 %

Table 3: Energy Performance at Avonmouth

The Thermal Hydrolysis (The Cambi Process)

They involve injecting steam at high temperature and pressure to rupture cell and improve conversion of organic matter to biogas digestion process. The most common thermal process is provided by Cambi. Cambi THP is a high pressure steam pre-treatment for anaerobic digestion of municipal and industrial sludge. Applying THP technology results in double digester loading increased biogas production and a pathogen-free and stabilized biosolid product with increase dewaterability.

Typical Process conditions are 15-17% Dry Solids at 165 degree centigrade with high pressure of 6 bars for less than one hour. The cooled 10-12 % diluted sludge is fed to a Mesophic Anaerobic Digestion (MAD) plant. The MAD process operates at a temperature of 42 degree centigrade. Biogas generated is used to produce CHP.

They involve injecting steam at high temperature and pressure to rupture cell and improve conversion of organic matter to biogas digestion process. The most common thermal Process is provided by Cambi. Cambi THP is a high pressure steam pre-treatment for anaerobic digestion of municipal and industrial sludge.

THP process uses 15-17% dry Solids at 165 degree centigrade with high pressure of 6 bars for less than one hour. The cooled 10-12 % diluted sludge is fed to a Mesophic Anaerobic Digestion (MAD) plant. The MAD process operates at a temperature of 42 degree centigrade. Biogas generated is used to produce CHP.

Cambi's above plant was build by Black & Veatch Contracting (UK) at Celtic Anglian Water (CAW). This plant is expected to handle about 91,000 tonnes of dry solids per year. This digestion would serve a population of 3 M people. The processes biosolids can be applied to land as fertilizer. The plant would generate up to 10 Mega Watts of electricity. UK has set a target of 15% energy supplied by the national grid to come from renewable source by 2020.

The plant has been designed for a maximum throughput of 40,000 tds/annum and will generate up to 4.7 MW of electrical power for use on site at Bran Sands whilst at the same time producing a Class A biosolid product. Overall, this will reduce the whole Bran Sands site electricity usage by up to 50 percent.

The modification to the drying plant at Bran Sands Regional Sludge Treatment Centre utilizes existing liquid sludge storage facilities and dewatering facilities for raw indigenous sludges and for the final digested sludge product.

The new AD plant consists of a twin cake reception facility and associated cake blending system, a two stream Thermal Hydrolysis Process (THP), three new anaerobic digesters, two gas storage units, four gas engines and waste heat recovery units plus two package boilers and associated water treatment facilities.

A new cake storage facility will also allow seven days storage of digested cake on site. The project is nearing completion with all functional testing complete. Full process commissioning commenced in August 2009.

The first Cambi Thermal Hydrolysis plant in Australia is installed at Oxley Creek Sewage Treatment Plant (STP), Brisbane, as part of the Brisbane Water Environmental Alliance (BWEA) 's sludge treatment center. The Cambi process treats up to 95,000 tonnes of dewatered waste activated sludge (WAS) from the Oxley Creek STP and trucked in from outlying STPs.The sludge is initially dewatered to 13.5% dry solids, then pumped sequentially into the pulper/preheater tank, where it is heated with steam recycled from the process. It is then pumped into the high-pressure reactors, where hydrolysis at high temperature (155°C) and pressure (4.5 bars) occurs. Retention time in the reactors is 20-30 minutes, after which the pressure is reduced to 2-3 bars, and the sludge flashed by pressure differential into the "flashtank".

The surplus steam is recycled to the preheater tank. The sludge is then pumped through a heat exchanger, into the digesters, at 8% dry solids. The Cambi process causes hydrolysis of the microbial cells and the sludge particles through a combination of temperature and sudden pressure reduction. Process gases are trapped, and injected into the anaerobic digester feed, eliminating any odors.

The Cambi process has complete duty and standby on all rotating equipment. Construction commenced in December 2005, and the plant was mechanically completed in March 2006.BWEA tested a number of disintegration and hydrolysis technologies. Cambi was the only one that could guarantee significant biogas production and good dewatering.

Grade A biosolids

12,900 tonnes DS/year	1,000 kW electricity
13,500 m³/day biogas	43 % DS reduction
Final product: 30% DS	50% VS destruction

Table 4: Plant output data.

Summary Analysis of Using Therophilic Digestion

This process ranked highest in terms of costs, reliability, and flexibility. It also performs optimally under stress conditions such as fluctuating flows and "Shock" loads.

Summary Analysis of Using WELtec Bio Power

WelTec provides 'Advanced single stage digesters'. The single stage process is made possible by the automation and fine tuning of the mixing input and output sequences made possible by the computer control of the pumps and stirrers in the digester. Probably the most significant advantage is the factory fabrication of stainless steel as against the conventional concrete construction. Other advantages are:

- Build time is fast
- Impact on the build site is reduced.
- Less weather dependent
- Longer life
- The stainless steel has a residual value.

Summary Analysis of Using Monsal System

In order to highlight the benefits of the Monsal System I would highlight one of their latest project at KINGS LYNN, undertaking mesophilic and thermophilic duties.These plants can treat up to 19,000 tonnes Dry Solids per year. The plants treat a variable load of 25 tonnes per day. Biogas production rate is 400m3/tonnes DS treated compared with 300m3/tonnes DS produced using conventional method, typically 30% uplift.

The percentage of VS destruction is at 60% conversion. There are now 11 full scale Monsal plants using this technology in U.K for Anglian Water, Southern, United Utilities, Wessex and Welsh Water.

Summary Analysis of Cambi Process

- Increased digester capacity for imported sludge (doubling through put of the existing digesters and making capacity for next 20 years.
- Improvement of final product to a Class A well dewatered product
- 43% dry solids reduction in the digesters
- 30% dry solids in the final cake product
- Increased loading rate of digester at ~ 8% DS
- Positive energy balance, i.e. only 25% of the biogas is used directly to operate the system. The remaining 75% is used to generate green electricity and process steam

Compare and Contrast Cambi vs. Monsal

Enzymic Hydrolysis and thermal hydrolysis have emerged as the fastest growing and most proven pre-treatment technologies in the UK. Each system has advantages and disadvantages over each other. One of the key differences between the two is the energy balance.

The key to the AAD process is a phase that significantly enhances the breakdown of organic materials by breaking down the cell walls. With thermal hydrolysis this is achieved by an initial high temperature of 165 degree centigrade combined with high pressure of 6 Bar for less than one hour. With enzymic hydrolysis this is achieved by phasing an increased temperature from 42 degree to 55 degree centigrade. The result is far greater conversion of organic matter into biogas when the material is transferred into thee anaerobic digestion phase.

Thermophilic digestion has many advantages such as higher metabolic rate and higher consequent specific growth rate compared with mesophilic digestion.

1. In contrast, Biological hydrolysis will generate more net energy in comparison to thermal hydrolysis. Biological hydrolysis was self-sufficient in energy.
2. Specific biogas yield per tonne of dry solid feed is comparable
3. EH and EEH is a biological system and is prone to upset with variable sludge feed—you cannot switch on and off as much as Cambi—important for sludge centres required in China.
4. EH and EEH are prone to E Coli reactivation on dewatering that gives doubt about meeting USEPA class A standard.
5. Most of pathogens are destroyed in the thermophilic anaerobic process which is effective against pathogenic bacteria which are inactivated within 24 hrs while weeks or even months will be needed under mesophilic conditions. In fact this is an important criterion for the municipal solid waste treatment, since the effluent can be used as a soil conditioner or fertilizer.
6. Generally EH is applied as a retrofit to plants that have large digestion capacity.

7. There have been a lot of failing EH and EEH digesters in China and I think process needs updating.
8. Central Sewage system is not sustainable in developed countries and can never be made sustainable in developing countries.

The effectiveness of these processes has become limited over the last two decades because of new Challenges of emerging environmental and regulatory issues of concern.

New Plant Implementation Criterion

To create a frame work of implantation plan the followings are needed.

- Creating Economic Framework
- Creating Regulation Frame work
- Minimizing nuisance including odor and elimination of pathogen.
- Creating Instrumentation Framework to monitor performances.
- Quality improvement of the waste water
- Quality of nutrient recovery (elimination of pathogen)
- Minimizing odor nuisance
- Building Capacity (Tripartite Poly System with Stake holder's involvement_ Government/ Business/ Public)
- Supply chain value creation
- Training and information sharing

Waste Water Treatment in USA

The EPA under President Barack Obama has set aside about one-third, or $3.3 billion, of its proposed fiscal 2011 budget for drinking and wastewater projects, almost double the dollar value approved in the final year of the Bush administration. Jackson said at the Association of Metropolitan Water Agencies' conference in Washington, according to the statement. "That means fostering innovation that can increase cost-effective protection."

Waste Water Treatment in UK

Every day in the UK about 347,000 kilometers of sewers collect over 11 billion liters of waste water. This is treated at about 9,000 sewage treatment works before the treated effluent is discharged to inland waters, estuaries and the sea.

There are more than 20,000 (combined sewer overflows) CSOs around the UK, all owned and operated by UK water companies. They are meant to act as safety valves for the system during periods of intense and heavy rainfall. They combine storm water with raw sewage and spill out of CSOs into rivers and eventually into the sea.

Since the early 1990s—with an estimated £8bn spent by water companies over the past 20 years—as Britain worked to comply with the European water directives that dictate acceptable water-quality levels. The bigger concern is that the general water quality testing system approved by the Environmental Agencies (EA) does not take the impact of CSOs into account up to date.

Waste Water Treatment in China

It is estimated that only a quarter of all wastewater in China is treated. Some progress has been made in recent years, but mainly with industrial wastewater, where official figures suggest a 90 per cent treatment rate on average in urban areas. However, in most rural regions, domestic wastewater is not treated at all.

According to China Water Website, only 1,572 sewage plans are working and 2,063 sewage plants under construction. It's estimated that 175,000 tons of sludge per day are generated. In order to treat 50 per cent of the wastewater from domestic households, 10,000 plants would be needed nationwide.

The direct discharge of wastewater into rivers and lakes is the main cause of the country's poor water quality in the first place. The situation is compounded by over-fertilization and excessive use of pesticides in agriculture and the leakage of industrial or domestic wastewater through defective sewage pipes.

The protection of rivers and lakes is a great environmental and social challenge in China. Demanding state-defined standards in wastewater treatment are forcing wastewater treatment plants to upgrade their technology. Chinese Ministry of the Environment (MEPA) in 2008 extended the regulation to highest stage of waste water quality. Some of the more developed provinces, like Jiangsu, have realized the danger of the sludge and have put forth further regulations and incentives of their own.

According to the eleventh five-year plan, the government aims for investment in sludge treatment to reach RMB 47.3 billion. Sludge treatment is also likely to receive some of the RMB 4 trillion government stimulus plans

China is building new socialist villages in its current 11th five-year plan. Developing biogas fits in well with its programme. China is reportedly planning to produce 15% of its electricity from low-carbon technology by 2020, and reduce its carbon density by 40 % from the 2005 level.

Obama is aware that "the nation that leads the clean technology will be the nation that leads the global economy".

In addition foreign capital and know-how will also be needed. Foreign companies have been active in the Chinese water and sanitation market for some time now. Build, operate, and transfer (BOT) projects are a common model, particularly in urban areas. Co-operation agreements and joint ventures with local companies are other possibilities.

Petroleum from Biomass

What is biomass?

Biomass is a type of fuel that produces energy by burning plant and animal matter such as trees, animal by-products, and general waste products. Wood remains the most commonly used biomass fuel, and is used to power systems that typically convert the energy into heat for water and space heating.

What are the benefits of biomass?

In terms of both installation and running costs, biomass has the lowest capital cost of all renewable energy technologies. Wood fuel is also a carbon neutral resource and can make a significant contribution to meeting the UK's commitment to reducing CO_2 emissions.

What is Bio fuel?

Biofuel typically refers to feed stock which produces Biogas or combined Heat and Power. Gas produced by the anaerobic digestion or fermentation of organic matter including manure, sewage sludge, municipal solid waste, biodegradable waste or any other biodegradable feedstock, under anaerobic conditions. Biogas is comprised primarily of methane and carbon dioxide. Depending on where it is produced, biogas is also called: swamp gas, marsh gas, landfill gas, digester gas. Biogas can be used as a vehicle fuel or for generating electricity. It can also be burned directly for cooking, heating, lighting, process heat and absorption refrigeration.

Non-Food Biomass

World has already witnessed food crisis due to unthought for quest to provide biofuel. Extra demand for agricultural produce, the food price has grown up to 75% in certain regions of developing countries. Massive expansion of biofuel production in the US has displaced food crops and taken land out of food production, contributing to a massive hike in food prices worldwide. Biomass power projects face the risk of shortage of feedstock. Since it is a folly to grow plants for the purpose of generating biomass, on farm lands where other crops can be grown. Recently on 23rd October1910, the Financial Times reported that corn prices had the biggest surge in prices since 1973 as expectation of drastic shortfall in crop production.

East European countries like Kazakhstan when exports grain for food; they get $120 per ton. But if they export grain for ethanol, they get $650 per ton. What is the better deal? If the ethanol can then be converted to ethanol oxide, this can then be sold at $1,100 per ton. Best deal would be if they produce ethanol themselves which they are doing. Biohim Co. is now producing at Kazakhstan ethanol from wheat and has a capacity of 60,000 tons/ year.

Biofuel Source Development

1st generation produced by fermenting sewage and 'food' crops such as Sugar cane, corn, wheat, and sugar beet.

2nd generation biofuels are produced by using biomass consisting of the residual non-food parts of current crops, such as stems, leaves and husks that are left behind once the food crop has been extracted. Other crops that are not used for food purposes (non food crops), such as switch grass, jatropha, sorghum, industry waste such as woodchips, skins and pulp from fruit pressing.

On February 8, 2010 Endeavour Space shuttle blasted off from Kennedy Space centre taking with it new scientific experiments, including a study of Jatropha curcas plant, used for producing biofuel, to see if its breeding process can be speeded up for commercial use.

Conversion Technologies enable the co-production of advanced biofuels, green power and other biogas products from organic wastes and hydrocarbons, taking the world beyond the use of food resources in the production of biofuels. The production of low-cost electricity and ethanol (and in the future, butanol or hydrogen) from waste streams will supplement gasoline, convert vast quantities of waste to energy and significantly reduce greenhouse gases, communities' costs of waste disposal, the need for landfills and the nation's dependence on foreign oil.

3rd generation biofuels are municipality waste using waste streams such as steam industrial waste such as CO2, "FOGS" (fats, oils, and greases).

No one technology is suitable for all waste streams. No single waste management system can handle the full array of waste resources. Each can form part of an integrated waste management system which is based on feed stock, treatment technologies and local market need.

Acid Catalyzed Organosolv Saccharification

This is the ACOS Technology patented by Paszner Technologies Inc. The process consists of a patented, super-dilute (<0.05%) acid catalyzed, total, single step simultaneous hydrolysis of both carbohydrates and lignin/extractives in a congruous low boiling organic solvent solution.

Solvent recovery/recycling separates the carbohydrates and lignin/extractives platform for direct refining into cellulose ethanol and other value added chemicals, food, pharmaceutical and cosmetic products.

The ACOS process operates on a closed water cycle and is water self-sufficient on processing green (. 50% moisture content) feed stocks. The process recovers 99.95% of the solvent and is solvent self sufficient. The ACOS process operates without wastes and pollution.

The ACOS Technology

Characteristics of the Acid Catalyzed Organosolv Saccharification is a process for producing ethanol-from-cellulose.

The process consists of a patented, super-dilute (<0.05%) acid catalyzed, total, single step simultaneous hydrolysis of both carbohydrates and lignin/extractives in a congruous low boiling organic solvent solution.

The ACOS process hydrolyzes random sized un-pretreated green biomass particles into sugars and lignin in a continuous, counter-current pulping reactor in less than 60 minutes and as little as 5 minutes depending on the reaction temperature used.

The ACOS process converts all lignocelluloses materials: coniferous, deciduous and agricultural residues with equal efficiency.

- Mixed feedstock processing is normal.
- Ethanol recovery is 380 L/T (100 gallons/T) from coniferous woods, 350 L/T (92.5 gallons/T) from deciduous woods and straws and 280—310 L/T (74 to 82 gallons/T) from cornstover and bagasse, respectively.

Summary of the ACOS Technology Benefits

The ACOS process can produce ethanol from biomass at small economies of scale and thereby, is suitable for regionally distributed ethanol production to reduce feedstock and fuel product transportation costs.

- ACOS processing economics does not depend on the ethanol value.
- The ACOS process refines biomass to value-added co-products to maximize the chemical value recovery and minimizes the production cost of ethanol.
- ACOS ethanol can be produced totally (>96%) carbon neutral in energy self-sufficient refineries.
- The ACOS process is the cleanest, least energy demanding and most efficient wood hydrolysis technology currently competing world wide.
- The process is an efficient job creator in both raw material procurement and refinery operating processes.

Enerkem's process

Enerkem's process combines green gasification and catalytic synthesis. It involves heat, pressure, advanced chemistry and the use of catalysts. Enerkem's extensive gas conditioning produces a tailored syngas that is converted first into methanol, then upgraded to ethyl acetate and then to ethanol.

Methanol is an ingredient used in a variety of compounds. Around 40 percent of methanol is converted to formaldehyde, and, from there, into products such as plastics, plywood, paints and textiles. Methanol is also used as a solvent and antifreeze, as well as a transportation fuel. According to SRI Consulting, world demand for methanol is projected to grow at an average annual rate of 7.8 percent from 2008 to 2013. Today, methanol is generally produced synthetically from natural gas. By converting residual biomass into methanol, Enerkem offers a unique product, biomethanol.

Acetic acid is an important chemical raw material. It is mainly used for the production of vinyl acetate monomer (VAM), but its fastest growing use is for its second largest derivative, purified terephthalic acid (PTA), which is driven by the demand in polyethylene terephthalate (PET) bottle resins and polyester fiber. According to SRI Consulting, Asia is expected to account for over 57 percent of acetic acid consumption in 2011 and the United States is expected to remain a major player, accounting for an estimated 19 percent of demand in 2011.

Acetates, such as ethyl acetate and methyl acetate, represent a large market. Ethyl acetate is used in a variety of coating formulations, such as epoxies, urethanes, cellulosic, acrylics and vinyls. Applications for these coatings are numerous, including wood furniture and fixtures, agricultural, construction and mining equipment, auto refinishing, maintenance and marine uses. Methyl acetate is mainly used as a chemical solvent for cleaning/coatings, and in its high-purity form, as a solvent for the pharmaceutical industry.

Enerkem's technology has been tested on more than 20 different feedstocks, such as wood chips or municipal waste, for more than 4,000 hours. Enerkem recently started construction in Edmonton, Alberta, on what it says is the world's first industrial-scale plant turning municipal waste into biofuel. Government policy, increasingly expensive and scarce landfill sites, and a growing demand for clean power are poised to drive a surge in demand for biofuel. In the United States, for example, government mandates call for production of 36 billion gallons of biofuel annually by 2022. Corn stocks will produce just 15 billion gallons. Edmonton will turn 100,000 tonnes of non-recyclable garbage each year into 36 million liters (10 million gallons) of ethanol, under a 25-year contract that will boost the city's waste diversion rate to 90 percent from 60 percent.

Advantages

- Less capital intensive
- Profitable at lower operating scales
- Simpler to operate Feedstock flexible—not limited to clean homogeneous biomass Low technology
- All systems and catalysts are industrially proven.
- Can be self-sufficient in energy requirements (heat and electrical)
- Minimal water requirement & water disposal

According to earlier analyses of Enerkem's economics, the plant design is financially feasible at a zero feedstock cost, based on $80-$100 oil. In the case of the Edmonton project, the construction of the facility enables the city of Edmonton to avoid construction of a new city landfill, and the city has contracted to supply 100,000 tones of MSW to the plant for an undisclosed tipping fee.This will produce 36 million liters of biofuels a year and reduce Alberta's carbon dioxide (CO_2) footprint by six million tones over the next 25 years-the equivalent of removing 42,000 cars off the road every year.

The Fischer—Tropsch process

Fischer—Tropsch Synthesis is a set of chemical reactions that convert a mixture of carbon monoxide and hydrogen into liquid hydrocarbons. The process, a key component of gas to liquids technology, produces a petroleum substitute, typically from coal, natural gas, or biomass for use as synthetic lubrication oil and as synthetic fuel. The F-T process has received intermittent attention as a source of low-sulfur diesel fuel and to address the supply or cost of petroleum-derived hydrocarbons.

Generally, the Fischer—Tropsch process is operated in the temperature range of 150—300 °C (302—572 °F). Higher temperatures lead to faster reactions and higher conversion rates but also tend to favor methane production. As a result, the temperature is usually maintained at the low to middle part of the range. Increasing the pressure leads to higher conversion rates and also favors formation of long-chained alkenes both of which are desirable. Typical pressures range from one to several tens of atmospheres. Even higher pressures would be favorable, but the benefits may not justify the additional costs of high-pressure equipment.

The largest scale implementation of F-T technology are in a series of plants operated by Sasol in South Africa, a country with large coal reserves but lacking in oil. Sasol uses coal and now natural gas as feed stocks and produces a variety of synthetic petroleum products, including most of the country's diesel fuel.

One of the largest implementations of F-T technology is in Bintulu, Malaysia. This Shell facility converts natural gas into low-sulfur diesel and food-grade wax. The scale is 12,000 barrels per day.

The new LTFT facility scheduled to commission in 2010 at Ras Laffan, Qatar is based on the Sasol technology, using cobalt catalysts at 230 °C. It includes the "Dolphin Gas Project" plant, converting natural gas to petroleum liquids at a rate of 140,000 barrels/day, with additional production of 120,000 barrels of oil equivalent in natural gas liquids and ethane.

U.S. Air Force certification Syntroleum, a publicly traded US company (Nasdaq: SYNM) has produced over 400,000 gallons of diesel and jet fuel from the Fischer—Tropsch process using natural gas and coal at its demonstration plant near Tulsa, Oklahoma. Syntroleum is working to commercialize its licensed Fischer-Tropsch technology via coal-to-liquid plants in the US, China, and Germany, as well as gas-to-liquid plants internationally. Using natural gas as a feedstock, the ultra-clean, low sulfur fuel has been tested extensively by the U.S. Department of Energy and the U.S. Department of Transportation. Most recently, Syntroleum has been working with the U.S. Air Force to develop a synthetic jet fuel blend that will help the Air Force to reduce its dependence on imported petroleum. The Air Force, which is the U.S. military's largest user of fuel, began exploring alternative fuel sources in 1999. On December 15, 2006, a B-52 took off from Edwards AFB, California for the first time powered solely by a 50-50 blend of JP-8 and Syntroleum's FT fuel. The seven-hour flight test was considered a success. The goal of the flight test program is to qualify the fuel blend for fleet use on the service's B-52s, and then flight test and qualification on other aircraft. The test program concluded in 2007. This program is part of the Department of Defense Assured Fuel Initiative, an effort to develop secure domestic sources for the military energy needs. The Pentagon hopes to reduce its use of crude oil from foreign producers and obtain about half of its aviation fuel from alternative sources by 2016.[19] With the B-52 now approved to use the FT blend, the C-17 Globemaster III, the B-1B, and eventually every airframe in its inventory to use the fuel by 2011.

Carbon dioxide reuse

In 2009, chemists working for the U.S. Navy investigated Fischer-Tropsch for generating fuels, obtaining hydrogen by electrolysis of seawater. When combined with the dissolved carbon dioxide using a cobalt-based catalyst, this study produced mostly methane gas. However, when using an iron-based catalyst, it was possible to reduce the methane produced to 30 per cent with the rest being predominantly short-chain hydrocarbons. Further refining of the hydrocarbons produced applying solid acid catalysts, such as zeolites, can potentially lead to the production of kerosene-based jet fuel.

The abundance of CO_2 makes seawater an attractive alternative fuel source. Scientists at the U.S. Naval Research Laboratory stated that, "although the gas forms only a small proportion of air—around 0.04 per cent—ocean water contains about 140 times that concentration" Robert Dorner presented the findings of his work to the American

Chemical Society on 16 August 2009, at the Marriott Metro Center in Washington DC.

Geobactor

Scientists have experimented with a variety of bacteria, but one kind that environmental microbiologist Derek Lovley at the University of Massachusetts at Amherst and his colleagues have focused on is *Geobacter*, which is naturally found in many soils and sediments. These have the novel ability to directly transfer electrons to the surface of electrodes. This had led to the construction of microbial fuel cells that are superior to previously described microbial fuel cells.

Geobacter species that have subsequently been isolated from a diversity of soils and sediments provide a model for important iron transformations on modern earth and may explain geological phenomena, such as the massive accumulation of magnetite in ancient iron formations.

Geobacter species are also of interest because of their role in environmental restoration. For example, *Geobacter* species can destroy petroleum contaminants in polluted groundwater by oxidizing these compounds to harmless carbon dioxide. As understanding of the functioning of *Geobacter* species has improved it has been possible to use this information to modify environmental conditions in order to accelerate the rate of contaminant degradation. As outlined under the Bioremediation link, *Geobacter* species are also useful for removing radioactive metal contaminants from groundwater.

Geobacter species also have the ability to transfer electrons on to the surface of electrodes. As outlined under the Microbial Fuel Cell link, this has made it possible to design novel microbial fuel cells which can efficiently convert waste organic matter and renewable biomass to electricity.

The genomes of several *Geobacter* species have been sequenced and are being incorporated into a computer model that can predict *Geobacter* metabolism under different environmental conditions. This systems biology approach is greatly accelerating the understanding of how *Geobacter* species function and the optimization of bioremediation and energy harvesting applications.

The MixAlco process utilizes naturally occurring microbes to convert cellulose into chemical intermediates and fuels. The focus of the early work was to identify organisms that utilize cellulose as an energy source, and then to select microbes that are hardy and efficient for use in industrial fermentation. Unlike a conventional cellulosic ethanol process, the MixAlco Process does not add enzymes to break down cellulose. Cattle, for instance, host a mixture of microbes that produce their own enzymes that break down cellulose and convert it into primarily acetic acid. Think of a cow's stomach as an industrial reactor, and you start to get the picture of what the process is attempting to accomplish.

From an economic point of view, the process has several advantages over conventional cellulosic ethanol. First, as noted, the process requires no enzyme addition. Second, because a mixed culture of hardy microbes is used, the process does not require sterile conditions. As you might imagine, the rumen digestive system is not a sterile environment. Finally, the process ultimately yields mixed alcohols or hydrocarbons, which have a higher energy content than does just ethanol.

Terrabon's Mix Alco technology is an acid fermentation process. The process can be applied to sewage sludge, municipality waste, by-products and non food crops like sorghum. It converts biomass into organic salts called carboxylic acids, which then can be converted to ketons and then refined through existing, conventional processes to produce gasoline, jet fuel or diesel.

From an economic point of view, the process has several advantages over conventional cellulosic ethanol. First, as noted, the process requires no enzyme addition. Second, because a mixed culture of hardy microbes is used, the process does not require sterile conditions. As you might imagine, the rumen digestive system is not a sterile environment. Finally, the process ultimately yields mixed alcohols or hydrocarbons, which have a higher energy content than does just ethanol. The reported yields are Mixed alcohols = 141 gal/tone.

Co-Production of Advanced Biofuel

Co-generation and Co-production are the conversion technologies of advanced biofuels and other biogas products from organic wastes and hydrocarbons. Just from 35.5 million tons of post recycled waste, conversion technologies could produce an estimated 1.6 billion gallons of ethanol and 12 MW green power. Ethyl Acetate and acetic acid are two in-demand chemical which are chemically quite close to ethanol. Base chemical ethylene is used to make many of our basic bioplastics and chemicals.

Honeywell and Rentech are building a co-production facility that will produce high amounts of fuel and electricity by 2012.They would provide eight airline companies at the Los Angeles airport to provide them with 1.5 million gallons of diesel produced from waste biomass. The company produces hydrocarbon fuels by gasifying biomass and utilizing the Fischer-Tropsch process to convert the syngas into liquid fuels.

As with a conventional petroleum refinery, each of these process steps in the bio forming process series can be optimized and modified to produce a particular set of desired hydrocarbon products. Gasoline products can be produced using the zeolite (ZSM-5) based process, jet fuel and diesel can be produced using a base catalyzed condensation route, and a high-octane fuel can be produced using a dehydration/ oligomerization route.

Municipality Waste-to-Fuels

Although landfill capacity remains insufficient throughout U.K as a result, higher tipping fees are turning attention to municipal solid waste ("MSW") as an attractive fuel feedstock. MSW includes everything from packaging, food scraps, and grass clippings, to old sofas, computers, tires, and refrigerators. The total amount of MSW generated throughout the country has tripled since 1960. While recycling and composting programs successfully diverted significant amount of waste away from landfills in 2007, the amount combusted or land filled still totals around 3 pounds per person per day.

MSW has many advantages over biofuels, which despite receiving generous support through subsidies and tax breaks in recent years, still faces significant obstacles to widespread commercial viability. Municipality Solid Waste can be transformed into Nation's most reliable domestic fuel source. MSW has a logistical advantage over traditional biofuels because it is cheap and abundant, already concentrated near population/demand centers, organized (collected, stored, and transported), and not subject to life-cycle analyses.

Educating the public about the difference between incinerators and waste-to-fuel production facilities to increase uptake going forward is important. The EU "policy package" is important to generate momentum for innovative waste solutions.

The EU policies include:

1. Banning green waste from landfills
2. Incentivizing technology for turning waste-to-fuel
3. Encouraging higher diversion rates
4. Holding producers responsible for the downstream disposal of products and packaging

Fats, oils, and greases were heralded as a significant resource for urban centers where a heavy concentration of restaurants can provide a steady source of biodiesel feedstock to power FOGS collection. The importance of civic engagement to facilitate collection processes and ensure that programs reach the critical mass to meet municipal biodiesel facility needs.

Regarding the subject of Valuation, Financing, and Implementation any process must demonstrate that the production facility will generate a profit. While carbon credits will translate into support for projects, they should only be treated as "icing on the cake" and not a guarantor of financial viability.

Most companies usually only have one piece of the puzzle: feedstock supply, an innovative technology, permitting, funding, citing, or a feasibility study. To succeed, pilot projects should seek partnerships to access the other pieces.

TMO Renewables

TMO Renewable developed a strain of "turbo-charged" bacteria that can turn tea bags, cardboard, wood and other household waste into fuel for cars and trucks. The technology is part of a wave of "second generation" biofuels that are not made from crop plants. TMO Renewable's bacteria converts rubbish and waste into biofuels. In 2008, TMO Renewables built the UK's first bioethanol plant that runs on grasses, cardboard and other waste. Key to its success was the incorporation of genetically engineered bacteria that can break down cellulose into simpler sugars, which can then be fermented to produce bioethanol. The TMO process relies on a strain of bacteria known as TM242 which grows at high temperatures of around 60C.

Fuel and Chemicals from Steel Mill Flue gas

LanzaTech has developed a proprietary gas fermentation technology for producing ethanol and high value chemicals (e.g.: MEK, Butadiene) from flue gas produced by a number of industries.

Any biomass resource (municipal waste, organic industrial waste (tyres), waste wood can be used as a nutrient source to produce syngas. The gasification process breaks down the chemical bonds in the biomass making up to 80% of the energy available for fermentation. The carbon monoxide containing gases are then scrubbed, cooled and sent to a bioreactor. The microbes use this energy to produce liquid biofuel. Once product recovery has taken place, the liquid biofuel produced can be used as a high octane premium fuel.

2, 3-BD can be readily converted to intermediaries like butenes, butadiene and methyl ethyl ketone that are used in the production of hydrocarbon fuels and a variety of chemicals including polymers, synthetic rubbers, plastics and textiles.

Algae are a large and diverse group of photosynthetic organisms. The algae are a paraphyletic group, meaning they did not descend from a common ancestor. Algal groups have independently evolved several times in life on Earth, representing a beneficial strategy adopted by parallel evolutionary paths.

The most complex algae of both the red and green versions, are called seaweeds, and can be found washed up on beaches around the world. Algae are the primary producers of the marine ecosystem and are consumed by a wide variety of organisms, especially filter feeders. Those animals that live deep in the oceans depend on dead algae or small animals floating down from above.

Compared to normal plants, microalgae are very efficient at converting light, water and carbon dioxide into biomass through photosynthesis. Algae are the fastest growing plants on earth and can multiply up to 2-3 times a day by cell division. Microalgae do not have roots, trunk, leaves or stems. That moves nutrients through the organism. Even though algae are representing only 30 percent of total plant biomass, they may create as much as 50 percent of the earth's oxygen.

Algae is serving an increasing important role as an emerging solution to create drop-in replacement fuels such as renewable diesel, renewable gasoline, and biojet fuel for commercial aviation, government and defense markets that are compatible with existing aviation, truck fleet, pipeline and storage infrastructure.

- Biofuel that can completely replace fossil fuels
- The yields of oil and fuels from algae are much higher (10-100 times) than competing energy crops has a high sugar content for conversion to ethanol and advanced biofuels, absorbs more airborne carbon than land-based plants, has no lignin, can be easily harvested compared to microalgae,
- Algae can grow practically anywhere, thus ensuring that there is no competition with food crops.

- Algae are excellent bioremediation agents—they have the potential to absorb massive amounts of CO_2 and can play an important role in sewage and wastewater treatment.
- Algae are already being used in a wide variety of industries and applications, and many newer applications are being discovered. Such a wide range of end-uses enable companies to produce both fuels and non-fuel products from the same algae feedstock
- Requires no pretreatment for ethanol production, and can be harvested up to six times a year in warm climates.

Macro Vs Micro Algae

- Microalgae have high oil content but are difficult to cultivate and harvest in a
- Cost-efficient manner.
- Macro algae, on the other hand, present low-cost cultivation and harvesting
- Possibilities, but most species are low in lipids as well as carbohydrates
- With processes such as cellulosic fermentation (for deriving ethanol), gasification
- (For deriving biodiesel, ethanol and a wide range of hydrocarbons), or anaerobic digestion (for methane or electricity generation), it is possible today to use
- Microalgae as the feedstock for biofuels.
- Thus, both micro and macro algae are potential feedstock for biofuels.

Bio Architecture Lab (BAL) and Statoil are jointly developing the technology and process to convert Norwegian seaweed into ethanol. Statoil is responsible for developing and managing the seaweed aqua farming operations, with consultation from BAL, which already has aqua farming operations in Chile.

Algae to Biodiesel

Biodiesel refers to a vegetable oil or animal fat based diesel fuel consisting of long chain alkyl is typically made by chemically these oils with an alcohol. Biodiesel is meant to be used in standard diesel engines and is thus distinct from the vegetable and waste oils used to fuel converted diesel engines. Biodiesel can be used alone, or blended with petrodiesel.

The President calls for five to ten commercial demonstration projects to be up and running by 2016.President Obama said, "Now, I happen to believe that we should pass a comprehensive energy and climate bill. It will make clean energy the profitable kind of energy, and the decision by other nations to do this is already giving their businesses a leg up on developing clean energy jobs and technologies. But even if you disagree on the threat posed by climate change, investing in clean energy jobs and businesses is still the right thing to do for our economy. Reducing our dependence on foreign oil is still the right thing to do for our security. We can't afford to spin our wheels while the rest of the world speeds ahead."

"Advancing biomass and biofuel production holds the potential to create green jobs, which is one of the many ways the Obama Administration is working to rebuild and revitalize rural America," said Agriculture Secretary Tom Vilsack. "Facilities that produce renewable fuel from biomass have to be designed, built and operated. Additionally, BCAP will stimulate biomass production and that will benefit producers and provide the materials necessary to generate clean energy and reduce carbon pollution."

Renewable Fuels Standard. EPA has finalized a rule implementing the long-term renewable fuels mandate of 36 billion gallons by 2022 established by Congress. The Renewable Fuels Standard requires biofuels production to grow from last year's 11.1 billion gallons to 36 billion gallons in 2022, with 21 billion gallons to come from advanced biofuels.

Since its beginnings in 2003, Solayzme has produced the world's first algal-based renewable diesel and the world's first 100% algal-based jet fuel. The company pioneered an unusual process, in that it makes algae indoors, without sunlight. The company claims that their algae are 1000 times more efficient at producing oils from sugar compared to growth by sunlight. Distillation is an energy-intensive process, and by using cellulosic-ethanol processing—using sugars that are not part of the human food supply—with their algae processing, they avoid the use of fuels required for conversion and distillation of alcohol-based fuel.

Solazyme's renewable oil production technology uses indirect photosynthesis bioproduction process uses microalgae to convert biomass directly into oil and other biomaterials, a process that can be performed in standard commercial fermentation facilities cleanly, quickly, and at low cost and large scale. They produce biofuel, plant oils and compounds in a diverse range of products from oleochemicals to cosmetics and foods.

Most macro-algae biofuel related projects prior to 2010 used on ethanol. Since 2010, the entrance of oil and petrochemical majors DuPont, Statoil and ENAP are expressing an increased interest in extracting sugars from seaweed to create not just ethanol, but also drop-in fuels, biochemicals and other valuable co products such as biobutanol and oleo chemicals. This follows a key trend by Shell and BP investing $12 and $8 billion respectively in sugar-based conglomerates in Brazil to produce ethanol, bio-butanol, drop-in fuels, and bio-based chemical products.

In September 2010, Bunge and Chevron invested in US-based Solazyme to create renewable algae-based oils using sugar as a feedstock. In addition, LS9, Amyris, and Virent aim to use plant-based sugars to produce bio-based diesel, drop-in fuels, biogasoline, biojet, biobutanol, biochemicals and bioplastics Will sea-based sugars from macro-algae provide a new feedstock for advanced biofuels, drop in fuels and biochemicals for these emerging sugar-based, infrastructure compatible biofuels and chemicals platforms?

In 2009, a field-to-wheels greenhouse gas life cycle test conducted by the Life Cycle Associates found that Solazyme's algal biofuel, Soladiesel™, emits 85 to 93 percent less GHG emissions than standard petroleum based ultra-low sulfur diesel. But not just that, it also found that its biofuels result in a significantly lower carbon footprint than any currently available first-generation biofuel as well.

Treatment Process and Nutrient Recovery

Supply Chain Management (SCM) over the past few decades has proven there to be opportunity to reduce cost and add value in supply chain. Traditional supply chain management has focused primarily on costs including transport, inventory and administration.

The recent emergence of sustainable supply chain management provides the opportunity to review processes, materials and operational concepts from a different perspective. It incorporates the role of the environment in supply chain value creation. The role of training and information sharing is most vital which we provide at CSBS.

Bioremediation for the breakdown of toxic pollutants to harmless product involves incineration. Resulting effect of the emissions, along with the high cost of supplemental fuel, making this a less attractive and less commonly constructed means of sludge treatment and disposal. There is no process which completely eliminates the requirements for disposal of biosolids. Incineration of waste materials converts the waste into ash, flue gas, and heat.

Nitrogen removal

The removal of nitrogen is effected through the biological oxidation of nitrogen from ammonia (nitrification) to nitrate, followed by de-nitrification, the reduction of nitrate to nitrogen gas. Nitrogen gas is released to the atmosphere and thus removed from the water. Nitrification itself is a two-step aerobic process, each step facilitated by a different type of bacteria.

Phosphorus removal

Phosphorus removal is important as it is a limiting nutrient for algae growth in many fresh water systems. (For a description of the negative effects of algae, it is important for water reuse systems where high phosphorus concentrations may lead to fouling of downstream equipment such as reverse osmosis.

Chlorination

Chlorination of residual organic material can generate chlorinated-organic compounds that may be harmful to the environment. Residual chlorine is toxic to aquatic species; the treated effluent must also be chemically dechlorinated, adding to the complexity and cost of treatment.

There are a number of established and emerging technologies with various applications such as Sulfate reduction for the removal and recovery of heavy metals and sulfur.

O & M

It is now essential to develop the O&M manual with all text and figures, specifically formatted for online delivery. This requires research, writing and illustrations as any O&M project. Special care should be taken in formatting and presentation. The O&M is then saved in the required formats. The materials can then be produced for effective online delivery.

Many people tend to forget the potential risks and liabilities associated with renewable energy. Health and safety needs to be considered and addressed from the earliest phase

Biofulels pose unique risks for handling and transportation. In addition other concerns are gaseous emissions, waste water—risk assessment and risk management—permission procedure, Liquid waste, Safety and health risks for the operators.

In developing countries following extra hazards should be considered when new plant/procedures are implemented. These are explosion,, fire, danger from electricity, poisoning, danger to health, harms to persons (burn, scald, etc.), irritation (skin, mucous membrane), mechanic failure, noise pollution, ototoxic effects, emergency stop gas engine, failure in combustion system, failure flare / emergency case gas utilization, failure/malfunction of the automation system.

Neat U Loo

In the development of any recognized world class eco-city, a dual system of water supply should be built in. Water for sanitation purposes need not use drinking water as feed stock. The separate sanitation system would eliminate industrial pollutants and the cost of recycling would therefore be much less. Savings should repay the installation cost within 10 years. Sanitation problem for developing countries can only be solved by adapting a poly system as being suggested here.

Neat U Loo is a community—Driven approach. It's a logic model. Its "needs led" and outcome focused. It asks what are the social concerns requiring attention. What are the outcomes that are right? Are the processes planned linked to the needed outcome? Are the outcomes are valued by the clients, beneficiaries, governmental organizations, Industry investors, Environmental friendly.

Neat U Loo is a Polysystem Hub. Polysystem is a hub for system and service which in collaboration with the stake holders create value chain of Bio-Energy. This Co-branding strategy introduces bargaining power by all parties with three lead members. What is being proposed is a dynamic poly system hub where users get paid.

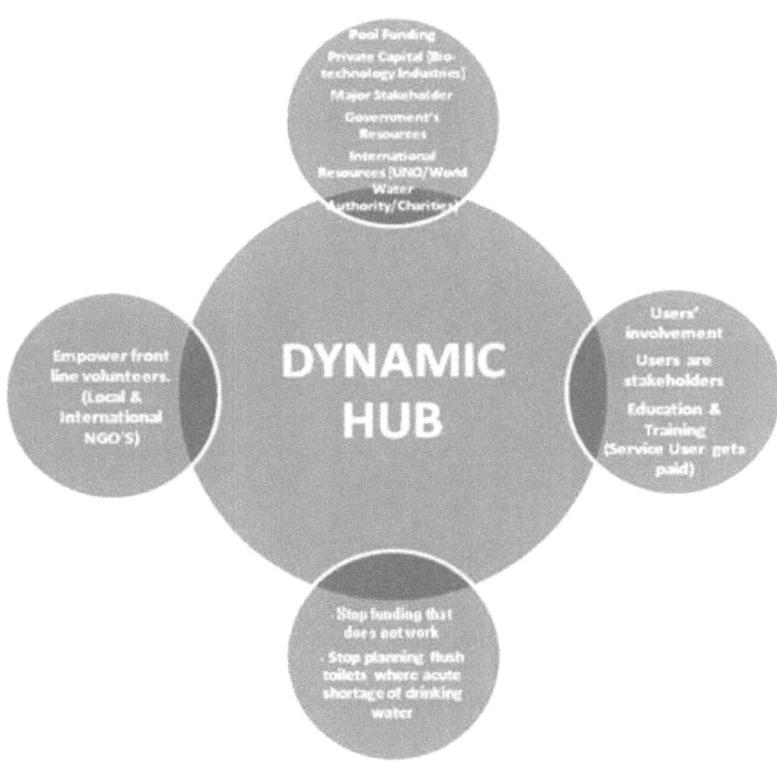

Neat U Loo Process Flow

Tripartite Partmership

Key stake holders (Lead Members) each must have separate and independent powers (fusion of powers) and areas of responsibility. The normal division of branches is into an executive, a legislature and a financial.

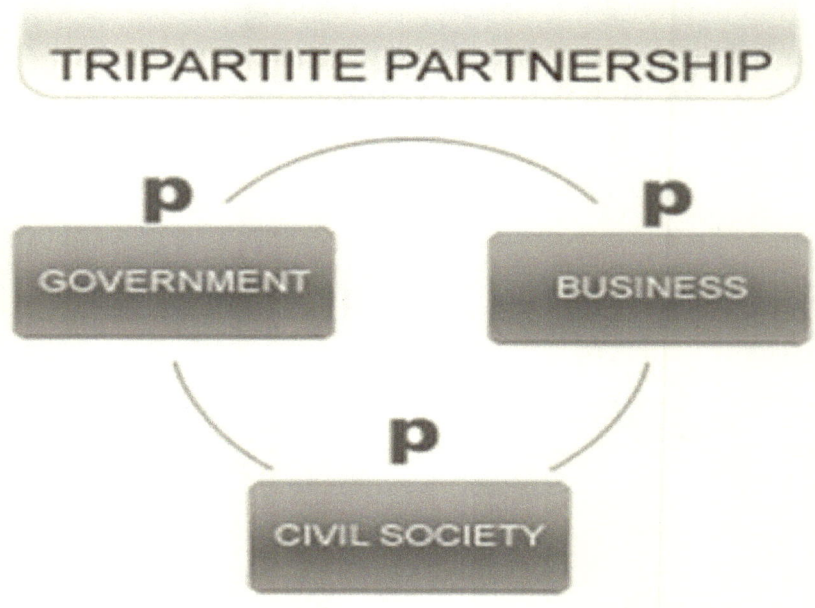

Tripartite Partnership

"Neat U Loo" Process Flow

In the development of any recognized world class eco-city, a dual system of water supply should be built in. Water for sanitation purposes need not use drinking water as feed stock. The separate sanitation system would eliminate industrial pollutants and the cost of recycling would therefore be much less. Savings should repay the installation cost within 10 years. Sanitation problem for developing countries can only be solved by adapting a poly system as being suggested here.

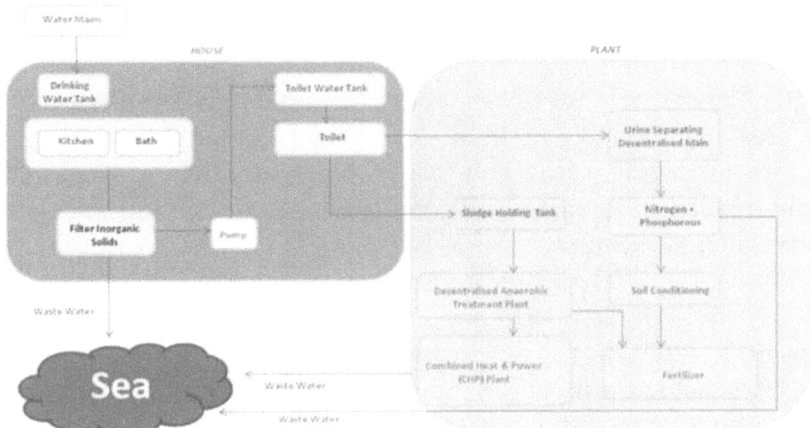

Parallel System (In and Out)

We have gained specialist knowledge of recent development in the field of Advanced Anaerobic Digestion processes is in close contact with some of the world leading contract designers and plant manufacturers. We can undertake early stage biotechnology operating consultancy work and negotiate cost effective innovative, proven processes.

Discussion

The main UK policies in favor of anaerobic digestion are the climate change levy and new renewable Obligation.

No one technology is suitable for all waste streams. No single waste management system can handle the full array of waste resources. Each can form part of an integrated waste management system which is based on feed stock, treatment technologies and local market need.

The switch to second generation biofuels will reduce competition with grain for food and feed, and allow the utilization of materials like straw which would otherwise go to waste. The biorefineries will also be able to use lignocellulosic crops like poplar and switch grass, which can be grown on land less suitable for farming than traditional row crops.

These findings should be a boost to companies hoping to establish themselves in this emerging field.

High oil prices, competing demands between foods and other biofuel sources, and the world food crisis, have ignited interest in alga culture for making vegetable oil, biodiesel,bioethanol, biogasoline, biomethanol, biobutanol and other biofuels, using land that is not suitable for agriculture.

The macro and micro algal populations of the aquatic environments provide a vast genetic resource and biodiversity. This feature alone suggests that these organisms have considerable potential for offering new chemicals, materials and bioactive compounds. The completion of the genome sequencing programmes of two microalgae also opens up major opportunities for new applications, either using the algae themselves or through using the genes in other production systems, whether fermenter-based or fields.

Researchers throughout the world are working to produce biofuel from algae. But a few are trying a decidedly novel approach: Using an abundant and freely available source—human waste—to make the fuel of the future while also treating sewage.

New Plant Implementation Criterion

To create a frame work of implantation plan the followings are needed.

- Creating Economic Framework
- Creating Regulation Frame work
- Minimizing nuisance including odor and elimination of pathogen.
- Creating Instrumentation Framework to monitor performances.
- Quality improvement of the waste water
- Quality of nutrient recovery (elimination of pathogen)
- Minimizing odor nuisance

- Environmental concerns of new technologies must not only consider carbon cycles and renewability issues, but also need to be operated safely to protect the immediate environment: fauna and flora, as well as, of course, residents and workers.

Educating the public about the difference between incinerators and waste-to-fuel production facilities to increase uptake going forward is important. The EU "policy package" is important to generate momentum for innovative waste solutions.

The EU policies include:

1. Banning green waste from landfills
2. Incentivizing technology for turning waste-to-fuel
3. Encouraging higher diversion rates
4. Holding producers responsible for the downstream disposal of products and packaging

Conclusion

Traditional supply chain management has focused primarily on costs. Through a combination of non-profit community based outreach voluntary organization and clean energy efficiency market place it is possible to design a simple cost effective sustainable supply chain. It is a win win win situation.

Bioenergy Technology hold promises, yet fail to deliver on reliability, cost and feed stock availability and quality.

Every energy production process comes with baggage of one sort or another. The key to successfully transitioning away from fossil fuels in the most sustainable manner will be to develop technologies in which the excess baggage is minimized.

Regarding the subject of *Valuation, Financing, and Implementation any process* must demonstrate that the production facility will generate a profit. While carbon credits will translate into support for projects, they should only be treated as "icing on the cake" and not a guarantor of financial viability.

In a conventional ethanol process (e.g., corn or sugarcane) a lot of the process energy is devoted to removal of water. There is also the issue that water resources are a problem in many areas

The end result could be very rewarding with proper selection of exact technology. After all this is the energy which is available today, it's available around the clock.

Energy that avoids the health related diseases.

The technology helps offset the production of harmful greenhouse gasses.

It has also been evidenced that thousands of non-bio-compatible substances move up the food chain.
The waste-to-fuel industry must neutralize public anxiety and strong opposition from environmental groups to get production facilities off the ground. Although policy is shifting increasingly in favor of waste-to-fuel solutions, a systemic overhaul of the waste management infrastructure is long overdue.

What is needed are sanitation systems that keep toxic and human wastes separates, prevent pollution, and return the nutrients in urine and faces to soil as fertilizers. Do not extend any sewer lines to presently unsewared houses, institutions or commercial facilities.

About a third of household water is used for flushing the WC. Greywater, the waste water from baths, showers and washbasins, can be collected in a household-scale reuse system and treated to a standard suitable for WC flushing.

More use should be made to use recycled water not requiring treatment to the level of drinking water. We must stop treating 10 liters of flushed Toilet water to get one liter of drinking water.

ACKNOWLEDGEMENTS

The author most sincerely thanks to Mr. Steve Bungay of Monsal, Mr. Peter Hough of Regreen & Co, Mr. Keith Panter of Ebcor Ltd,, and Mr. Herald Kleiven of Cambi. I would also like to take this opportunity to thank Dr. Son Le, MBE, United Utilities PLC, Dr. Jennifer Holmgren, CEO of Lanza Tech, Mackinnon Lawrence of Cleantech Law partners, Dr. Koenraad Vanhoutte, CEO at Navicula and founder of SBAE industries. Mr. Hamish Curran, CEO of TMO provided me with some valuable documents.I am immensely indebted to all of these brilliant people who have kindly engaged with me in detailed discussions ranging from plant details to theoretical process fundamentals. Lastly but not least I thank my wife Manjula and my son Sujoy who helped me all along with my dream of writing a second book.

REFERENCES

1. **Anaerobicdigestion**
    http://upload.wikimedia.org/wikimedia/Common/2/2b/
    Stages_of_anerobic_digestion.jpc

2. **Case Study: Anacis Island Wastewater Treatment Plant**
    Fortmann-Row, S.
    http://simgua.com/documents/case_study_wwtp.pdf

3. **"Weltec Anaerobic Digesters"**
    Hough,P
    http://www.regreen.co.uk

4. **"Operational Experience of advanced anaerobic digestion"**
    14th European Biosolids and Organic Resources Conference and Exhibition"
    Bungay,S.
    http://www.monsal.com/Documents/Technical%20
    Paper%20No%2045%20Bungay%20S%202009%20
    Operational%20Experience%20of%20Advanced%20
    Anaerobic%20Digestion.

5. **"CAMBI is awarded its largest Thermal Hydrolysis contract"**
    Panter, Keith
    http://www.cambi.no/wip4/detail.epl?cat=&id463450&l=1

6. **Cellulose EthanolRefining, Acid Catalyzed Organosolv Saccaharification**
    Paszner, Laszlo
    http://acos-biomas-refining.com/

7. **"Biofuel Pioneer Plans More Plants"**

Paxton; Robin,Kazakh
http://www.ecoearth.info/shared/reader/welcome.
aspx?linkid=64471

8. **"The lowdown on Enerkem's landmark Edmonton advanced biofue project"**
http://biofueldigest.com/bidigest/2010/09/01/the-lowdown-on-enerkems-landmark-edmonton-advanced-biofuels-project/

9. **"Geobactor Project"**
http://www.geobactor.org/

10. **"Renewable Energy Solutions"**
http://www.terrabon.com/

11. **"Waste-to-Fuels: Innovation, Cost Parity Offset by Public Anxiety . . ."**
Lawrence,M.
http://blog.cleantechies.com/2009/05/29/waste-to-fuels-innovation-cost-parity-offset-by-public-anxiety/

12. **"Develops and markets technology for the conversion of carbon-bearing materials into premium liquid hydrocarbons"**
http://www.rentechinc.com

13. **"The BioFormingprocess can speed the use of non-food plant sugars to replace petroleum as . . ."**
http://newenergyandfuel.com/http:/newenergyandfuel/com/2009/06/26/virent-is-the-bio-fuel-producer-to-beat/

14. **"TMO signs 20 year, multi-site bio-ethanol contract with Fiberight**
Curran, Hamish
www.tmo-group.com/news/news-20-09-10.aspx

15. **Difference between Micro and Macro Algae,**

http://www.algae.wur.nl/UK/factsonalgae/
difference_micro_macroalgae/

16. "Solayzme has produced the world's first algal-asedrenewable diesel . . ."
http://www.bioenergywiki.net/Aviation_Industry

17. Anaerobic Production of Magnetite by a Dissimilatory Iron-Reducing Microorganism
Lovley DR, Stolz JF, Nord GL Jr, Phillips EJP, 1987, Nature. 330(6145): 252-254.
http://www.*Geobacter*.org/refs/publications/Nature_1987_Nov.pdf

18. *Basic Science with an Applied Product*
http://www.*Geobacter*.org/

19. Biomass-to-fuel-via-the-mixalcoprocess
Rapier, Robert
http://www.consumerenergyreport.com/2010/07/19/

20. "Obama Announces Steps to Boost Biofuels, Clean Coal"
The White House, Feb 03, 2010
http://www.whitehouse.gov/the-press-office/obama-announces-steps-boost-biofuels-clean-coal

www.ingramcontent.com/pod-product-compliance
Lightning Source LLC
Chambersburg PA
CBHW050337290526
45785CB00006B/2524